# The Girl Down the Road

Dona Herweck Rice

**Publishing Credits**

Rachelle Cracchiolo, M.S.Ed., *Publisher*
Conni Medina, M.A.Ed., *Managing Editor*
Nika Fabienke, Ed.D., *Content Director*
Véronique Bos, *Creative Director*
Shaun N. Bernadou, *Art Director*
Carol Huey-Gatewood, M.A.Ed., *Editor*
John Leach, *Assistant Editor*
Courtney Roberson, *Senior Graphic Designer*

**Image Credits:** All images from iStock and/or Shutterstock.

**Teacher Created Materials**
5301 Oceanus Drive
Huntington Beach, CA 92649-1030
www.tcmpub.com

**ISBN 978-1-4938-9850-3**

Printed in China
Nordica.082018.CA21800936

We go down the 

road

to see a  .

friend

We go down the 

road

to see a  .

trampoline

We go down the 

road

to see a  .

swing

We go down the 

road

to see a  .

playhouse

We go down the 

road

to see a .

bicycle

We go down the 

road

to see a  .

scooter

We go down the 

road

to see a  .

hamster

We go down the 

road

to see a  .

bird

We go down the  road

to see a  sandbox.

We go down the 
road

to see a  .
pool

# High-Frequency Words

## New Words

down go

see we

## Review Words

a the

to